Erick Cardoso Costa
Rodrigo B. Nogueira

# Universal mechanical testing machine

**Erick Cardoso Costa**
**Rodrigo B. Nogueira**

# Universal mechanical testing machine

## Drive and control design

**ScienciaScripts**

**Imprint**

Any brand names and product names mentioned in this book are subject to trademark, brand or patent protection and are trademarks or registered trademarks of their respective holders. The use of brand names, product names, common names, trade names, product descriptions etc. even without a particular marking in this work is in no way to be construed to mean that such names may be regarded as unrestricted in respect of trademark and brand protection legislation and could thus be used by anyone.

Cover image: www.ingimage.com

This book is a translation from the original published under ISBN 978-620-2-18347-5.

Publisher:
Sciencia Scripts
is a trademark of
Dodo Books Indian Ocean Ltd. and OmniScriptum S.R.L publishing group

120 High Road, East Finchley, London, N2 9ED, United Kingdom
Str. Armeneasca 28/1, office 1, Chisinau MD-2012, Republic of Moldova, Europe
Printed at: see last page
**ISBN: 978-620-7-55900-8**

Copyright © Erick Cardoso Costa, Rodrigo B. Nogueira
Copyright © 2024 Dodo Books Indian Ocean Ltd. and OmniScriptum S.R.L publishing group

# Summary


This paper presents the design, dimensioning, manufacture and testing of a drive system for a universal mechanical testing machine. The system is used to generate loads in order to test polymeric material specimens. The project sought to develop a drive system that enables control and interaction with a microcomputer, featuring displacement accuracy and test speed variation. The methodology for developing the system consisted of studying machine drive models, structural calculations for sizing the load and the components used in the mechanical assembly, building the elements and testing the system's operation. The development of the drive system is discussed, from the design and choice of components to be used to the speed control system using the IDE. Tests of the drive system consisted of implementing the control code to analyze the system's operation, checking the performance of each mechanical component developed and the ability to control the stepper motor using the HY-DIV268N-5A driver and the Arduino platform. The results obtained will contribute to correcting the errors and inadequacies found in the system used and to developing the most suitable displacement and speed control for carrying out the tests using Arduino technology.

*Keywords:* equipment development, mechanical testing, Arduino technology, machine/equipment design

# TABLE OF CONTENTS

2

# 1. INTRODUCTION

Over the years, in order to increase the reliability and efficiency of structural projects, greater importance has been attached to the materials used as raw materials and, consequently, tests to evaluate and learn about the properties and mechanical behavior of materials have increased in order to save material and reduce weight and costs (REBOUÇAS, 2007).

The mechanical properties and behavior of the materials that make up structures under certain stress conditions are obtained through mechanical tests. These tests are used to assess the mechanical performance of structural parts of a wide variety of products, such as machines, buildings, etc. From these mechanical tests it is possible to obtain intrinsic material properties by analyzing the relationship between the stress generated and the deformation produced, making it possible to quantitatively express the mechanical properties required for application, product certification, quality control, comparison and selection of materials, research and development of new materials (BLANCO, 2006).

To carry out these tests, a mechanical system is usually used which subjects a sample of the material to mechanical stress (traction or compression) and, at the same time, measures the deformation resulting from the stress condition. The equipment commonly used is the universal mechanical testing machine which, by means of movement in the vertical direction of the mobile platform induced by a hydraulic, pneumatic or mechanical system, applies an axial load to the specimen which can be tensile or compressive, with variable speeds, which define the speed of the test (ASKELAND and PHULÉ, 2008).

To generate this load, there must be a system that drives the movement of the mobile platform. This system can be electromechanical, which will be described in section 2.1. A machine with an electromechanical drive generates a load through a system consisting basically of an electric motor, together with a set of reduction elements and one or more power screws for transmitting the load.

Technical standards define test speeds in ranges of values according to the type of material to be analyzed (LEMOS et al., 2013). Therefore, the electric motor must have variable speed and interact with a controller unit so that the torque and test speed parameters can be set via an operator interface.

This work is part of a series of projects aimed at developing a universal mechanical testing machine. The development of the machine is extremely complex as it involves various systems responsible for its operation (base structure, support, mobile platforms, side guides, motor coupling and electromechanical system) and accessories (load cell, grips and extensometer) needed to carry out the tests, so the overall design of the machine is being carried out in complementary stages.

Given the above, this work is justified by the importance of the electromechanical system for the development of the machine, since it will be possible to generate a load to carry out the mechanical tests.

This work aims to develop an electromechanical system that differs from existing machine system models on the market, as it will use differential elements such as: a single power screw, transmission assembly (pinion, chain and crown), threaded part for coupling and torque transmission, stepper motor, control driver and Arduino to control the stepper motor. Another differentiator is the cost-benefit ratio, in which the elements used to develop the system are easy to manufacture, functional, available for purchase and low cost when compared to equipment for this purpose available on the market.

The control system is another distinguishing feature of the project, which used the HY-DIV268N-5A driver connected to the simplified Arduino platform. This driver is low cost, small in size, easy to handle and easy to program, all of which are attractive features when it comes to automating equipment.

The design and manufacture of the testing machine at the Institute of Exact Sciences and Technology, in addition to bringing benefits to the Campus, will contribute to the development of technologies and innovation at the University and in the interior of Amazonas through the development of new technologies.

In addition to this introductory section with a contextualization of the topic and justification, this course conclusion is divided into:

- Objectives: the objectives of the project are described.

- Literature review: the main references on the subject under study are covered. Central concepts involving the development of a universal mechanical testing machine are discussed, such as: types of universal mechanical testing machines, tests carried out on the equipment, test specimens and the electromechanical system.

- Materials and Methods: this covers the procedure used to develop the drive system for the universal mechanical testing machine, divided into seven sections: load determination, torque transmission, support rod and plate separation, mechanical system assembly, motor and machine support and the control system.

- Results and Discussions: the results obtained from the development and testing of the proposed drive system are presented, analyzed and discussed.

Conclusion: the conclusions of the work are reported.

# 2. OBJECTIVES

## 2.1 General Objective

Developing an electromechanical system to drive a universal mechanical testing machine.

## 2.2 Specific objectives

    i.    Determine the load that the system must carry;

    ii.    Calculate the required gear ratio;

    iii.    Sizing the electric motor;

    iv.    Develop a torque transmission system;

    v.    Engage the motor;

    vi.    Develop a support for the machine;

    vii.    Develop the control system;

# 3. LITERATURE REVIEW

## 3.1 Universal mechanical testing machine

In the field of engineering, a vast knowledge of the characteristics, properties and behavior of materials is necessary to make structural projects feasible. Tests are a criterion for choosing materials, because through standardized methods, it is possible to know the mechanical properties and behaviour under stress conditions (GARCIA et al., 2013). The most appropriate way to determine the behavior of materials when subjected to stress loads is to carry out experiments in the laboratory. The usual procedure is to place small samples of the material in testing machines and apply loads to measure the resulting deformations (GERE, 2003).

There are two types of universal mechanical testing machine available on the market: hydraulic and electromechanical. Hydraulic machines work through the movement of a piston that drives the head to move vertically. To achieve precise speed control, it is necessary to adjust the orifice of a pressure-compensated needle valve to control the feed rate. Electromechanical machines operate on the basis of a variable speed electric motor, in which a system of reduction gears and one or more screws move the platform in a vertical direction. These movements make it possible to carry out tests with a variation in test speed by changing the motor speed modified by a controller unit (GRUPO CIMM, 2016).

The universal testing machine is widely used for its ability to carry out tensile, compressive and bending tests, as described in section 3.2. These tests are destructive, as they render the material useless (ASKELAND and PHULÉ, 2008).

## 3.2 Proof body

To carry out the tests, it is necessary to use a specimen, which is a sample of the material to be tested. It can be taken from the part (finished product) or from the material to be tested (SOUZA, 1982). To be used as a sample of the material to be studied, it is necessary to maintain a standard. To do this, standardized specimens are used according to specific technical standards for the type of material and the test to be carried out (ZOLIN, 2008). The American Society for Testing and Materials (ASTM), the Brazilian Association of Technical Standards (ABNT), the German Institute for Standardization (DIN) and the International Organization for Standardization (ISO) are examples of associations that determine the dimensions for making specimens. Figure 2 shows the typical shape of a specimen.

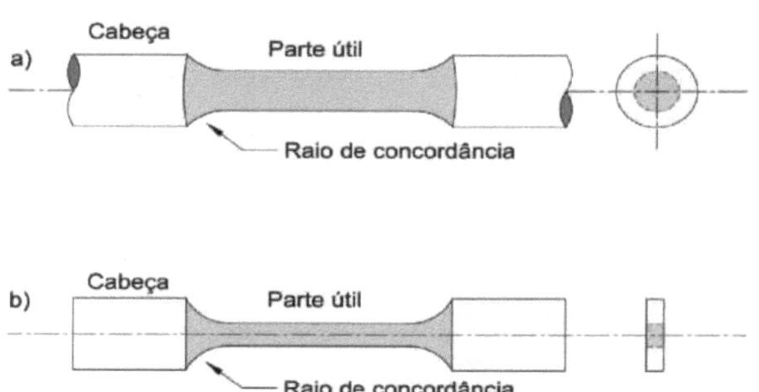


Figure 1: Specimen of (a) circular cross-section and (b) rectangular cross-section. Source: Adapted from Zolin (2008).

The specimen has a region called the useful part, which can have a rectangular or circular cross-section or depend on the geometry of the product being removed. The useful part is the area where the mechanical properties will be measured and the head of the specimen is used to fix it in the testing machine (SOUZA, 1982). The cross-section of the specimen will be circular (Figure 1a) for materials that are originally circular or have irregular shapes. Specimens obtained from plates, sheets, plates, films, fibers, tubes, wires, cables, etc., maintain the same cross-section as the product from which they originated (Figure 1b).

To carry out the mechanical test, the specimens are attached to the universal machine by mechanical clamps and an extensometer is used to measure the variation in their length, which is a device used to measure the deformation generated in the specimen when subjected to mechanical stress.

## 3.3 Tests carried out on the universal mechanical testing machine

The strength of a material depends on its ability to withstand a load without excessive deformation or rupture. This property is inherent in the material itself and must be determined by experimental methods. The most important tests in these cases are tensile, compressive and bending tests. These tests are mainly used to determine the relationship between stress and strain in many materials used in engineering, such as metals, ceramics, polymers and composites (HIBBELER, 2010).

To carry out the tests, a sample of the material called a specimen with a standardized shape and size is prepared. Two markings are made along the length of the specimen, as this is where the distribution of stress and strain will occur during the application of the load. The initial cross-sectional area of the specimen and the reference length, which is the distance between the two markings, as defined by technical standards, are measured. The ends are clamped and then the load is applied (HIBBELER, 2010).

Figure 2 illustrates the main types of mechanical tests and then describes them according to the type of load applied.

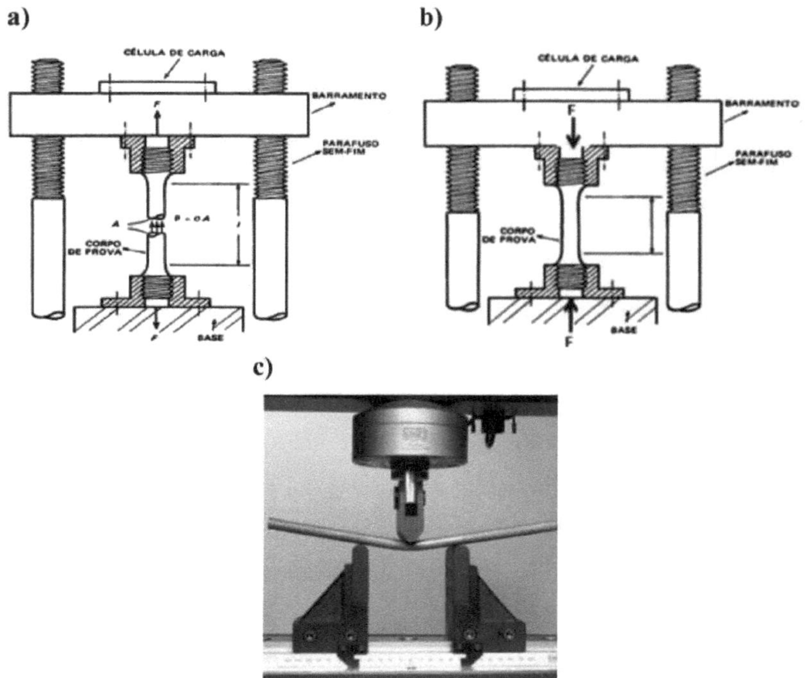


Figure 2: (a) Tensile test, (b) compression test and (c) bending test. Source: Padilha (2000) and EMIC (2015).

The tensile test consists of applying a uniaxial tensile load to a specimen until it breaks (Figure 2a) and at the same time measuring variations in its length (SOUZA, 1982).

The compression test consists of applying a uniaxial compression load to a specimen (Figure 2b). When a material is subjected to compression loading, the relationships between stress and strain are similar to those obtained in the tensile test (GARCIA et al., 2013).

In the bending test, bending is caused by applying a load at three points, causing a tensile stress that arises at the central and lower point of the specimen, Figure 2c (ASKELAND and PHULÉ, 2008).

## 3.4 Electromechanical system

In order to carry out the tests, it is necessary to move the platform with precision and speed variation (PESSÔA, 2014), so it is possible to make precise movements using a stepper motor. This equipment is widely used in industrial automation applications as it easily allows speed and rotation displacement to be controlled (AKIYAMA, 2009).

## 3.4.1 Stepper motor

The stepper motor is an electromagnetic actuator device that performs discrete angular movements through the interaction between electromagnetic fields. Its operating principle is the conversion of electrical pulses into mechanical movements following a digital logic (MONTEIRO, 2012; AKIYAMA, 2009).

With each pulse, the motor performs an angular rotation, i.e. the number of steps is exactly equal to the number of pulses received. Each step is a portion of the complete rotation. Thus, in order to advance a certain angle, several pulses are needed to reach the desired position. The stepper motor has 48, 100 or 200 steps to complete one revolution, which means angular steps of 7.5°, 3.6° or 1.8° respectively (PESSÔA, 2014).

The stepper motor has high speed, direction and angle control, and can be rotated at a specific angle with extreme precision, which is determined by the number of steps per rotation, i.e. the greater the number of steps, the greater the precision (VIVALDINI, 2009; ROMERO, 2010). Stepper motors can be unipolar or bipolar, and Table 1 shows the main differences.

Table 1: Main differences between unipolar and bipolar stepper motors. Source: Evans et al. (2013).

| Unipolar stepper motor | Bipolar stepper motor |
| --- | --- |
| Simpler to control | More efficient |
| Generally lower cost | Greater torque per unit of power |
| 5 or 6-wire connections | 4-wire connections |
|  | Higher rotation speed |
|  | Simpler construction |

The characteristic of this device is its torque x speed ratio (Figure 3) in which the speed of rotation of the shaft is a function of the time between steps, which is set by programming through a controller unit. The shorter the time between steps, the higher the angular velocity and the lower the torque provided (VIVALDINI, 2009).

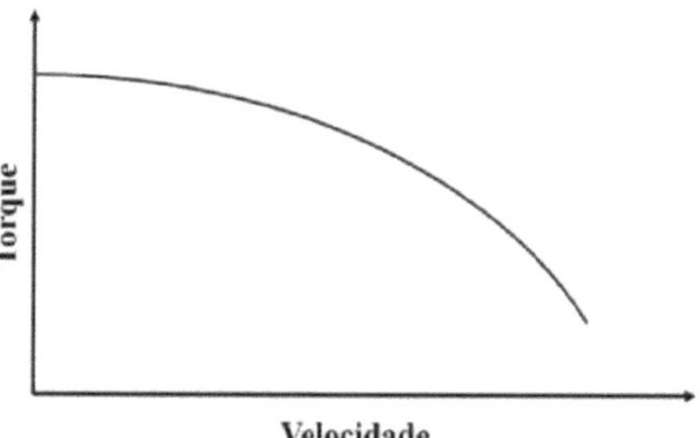


Figure 3: Torque x Speed. Source: Vivaldini (2009).

## 3.4.2 Arduino

Arduino is an open source platform based on a simple microcontroller board and a development environment for programming and processing inputs and outputs between the device and the external components connected to it. The *software* consists of a standard programming language and the *bootloader* (MULTILÓGICA,

2015).

The Integrated Development Environment (IDE) is open source and its programming is written in the language that arduino understands. The arduino *hardware* and *software* are open source, meaning they can be used freely (McROBERTS, 2011).

The arduino is based on ATMega8 and ATMega168, microcontrollers from Atmel. The plans for the modules are published under a *Creative Commons* license, so experienced circuit *designers* can make their own version of the module, extending and improving it (SILVEIRA, 2013).

There are several types of microcontrollers and microcontroller platforms available for physical computing (*Basic Stamp Parallax, NetMedia's BX-24, Phidgets, MIT's Handyboard*). However, arduino has advantages over other systems such as: low cost, as arduino boards have a relatively low acquisition cost compared to other platforms and microcontrollers; multiplatform, as arduino *software* runs on different operating systems; simple programming environment; open source extensible *hardware* and *software*, as arduino software is published with open source tools, available for extension by experienced programmers; has several board solutions: Uno, Mega, Nano, Diecimila, Mini, BT, Duemilanove, each with its own characteristics serving different purposes; wide field of action for controlling systems and can have applications in the area of 3D printing, robotics, engineering and others (WERNECK, 2013).

### 3.4.3 HY-DIV268N-5A Driver

The HY-DIV268N-5A Driver is a high-performance *microstepping* device suitable for driving 4, 5, 6 and 8-wire hybrid stepper motors with two phases and requiring a power supply voltage of 12 to 48V. It supports input current between 1 and 5 Amps, selectable via switch, and output current between 0.2 and 5 A (DATASHEET HY-DIV268N-5A).

It uses the high-efficiency Toshiba TB6600HG chip that adopts *single-chip* PWM to ensure low vibration and high efficiency. To guarantee a maximum output current of 5A and a 48V output withstand voltage, it uses BiCD0.13 process technology in the chipset, allowing it to operate 2-phase or 4-phase motors (DATASHEET HY-DIV268N-5A).

This device uses the unit's subdivision control providing low motor torque ripple, optimum performance at low speeds, almost no vibration and noise. Torques are higher than other two-phase drives and high positioning accuracy. It is widely used in engraving machines, CNC machine tools, packaging machines and other equipment requiring fast responses and speed and positioning accuracy.

### 3.4.4 Bearing

Bearings are mechanical components used to enable rotational or linear movement, improve the functioning of machines, save energy and reduce friction, thus increasing speed and efficiency. They are extremely important devices for the stable operation and guaranteed performance of machines. Figure 4 shows the main bearing components.

Figure 4: Bearing structure. Source: Koyo (2009).

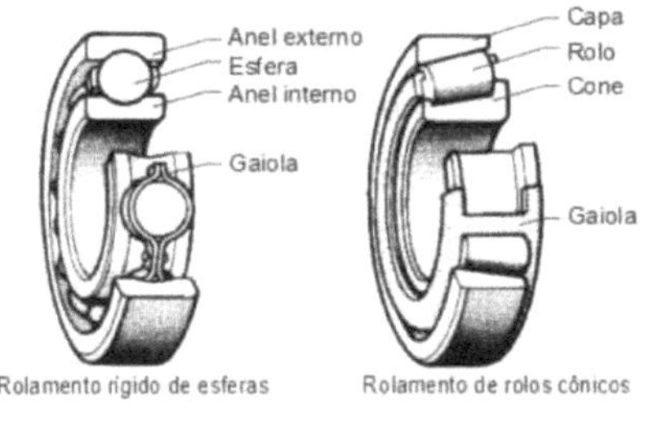


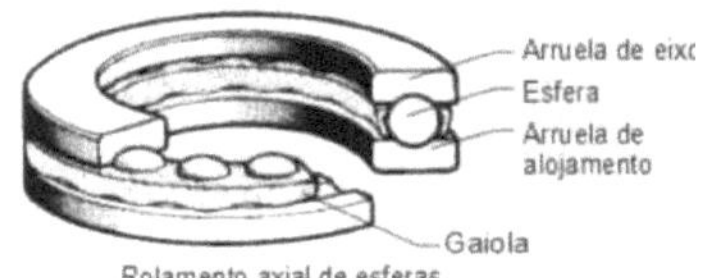


They consist of bearing rings, rolling elements and a cage. The rolling elements are arranged between the rings retained by a cage in the correct relative position so that they do not touch. This structure provides smooth rolling movement during operation (KOYO, 2009).

Bearing components have been developed and improved to perform the basic function of reducing mechanical friction and must fulfill a wide variety of needs, so they come in different shapes and varieties, as shown in Table 2.

Table 2: Main types of bearings. Source: NSK Brasil.

| | |
|---|---|
|  | **(1) Ball bearing:** widely used bearing. |
| 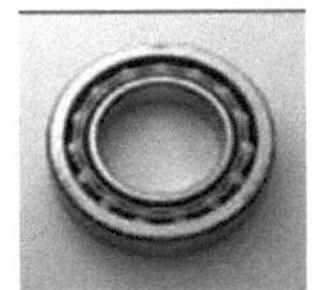 | **(2) Angular contact ball bearings:** this model has an angle of contact between the balls and the inner and outer ring. This equipment can withstand radial and axial loads. |

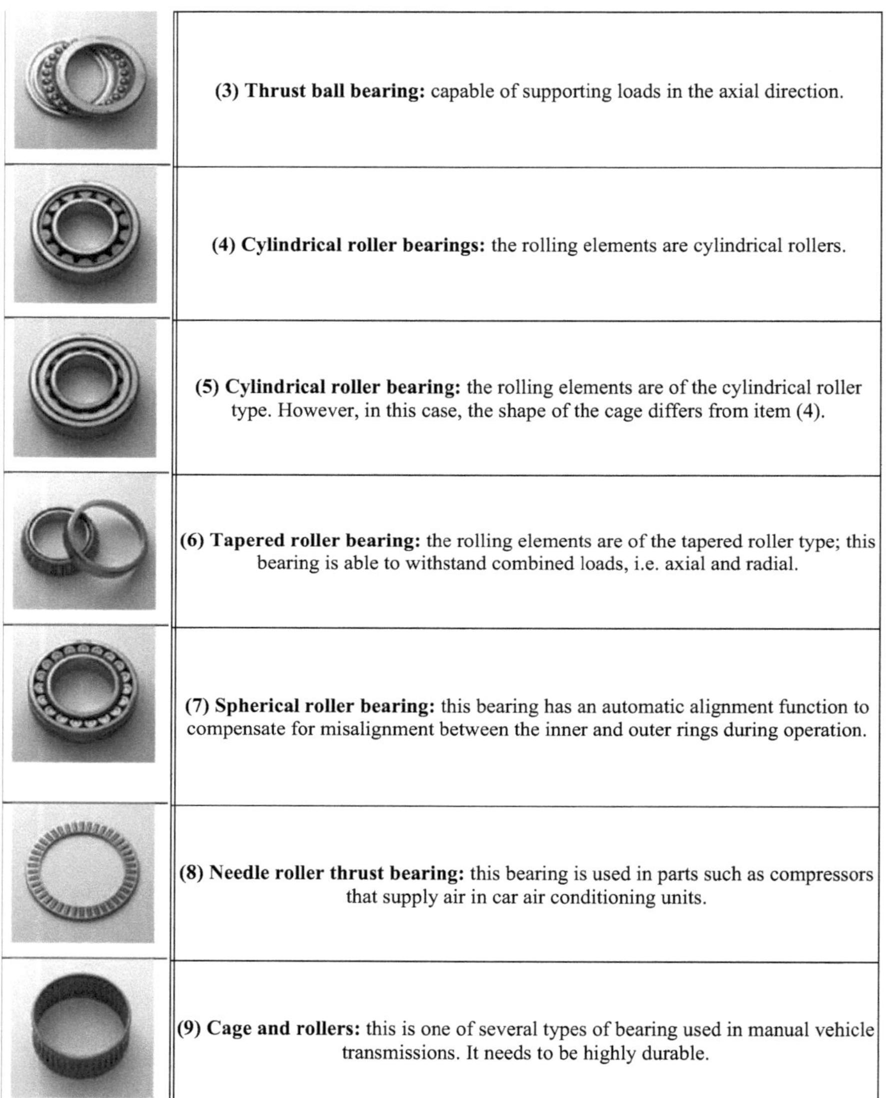

| | |
|---|---|
| | **(3) Thrust ball bearing:** capable of supporting loads in the axial direction. |
| | **(4) Cylindrical roller bearings:** the rolling elements are cylindrical rollers. |
| | **(5) Cylindrical roller bearing:** the rolling elements are of the cylindrical roller type. However, in this case, the shape of the cage differs from item (4). |
| | **(6) Tapered roller bearing:** the rolling elements are of the tapered roller type; this bearing is able to withstand combined loads, i.e. axial and radial. |
| | **(7) Spherical roller bearing:** this bearing has an automatic alignment function to compensate for misalignment between the inner and outer rings during operation. |
| | **(8) Needle roller thrust bearing:** this bearing is used in parts such as compressors that supply air in car air conditioning units. |
| | **(9) Cage and rollers:** this is one of several types of bearing used in manual vehicle transmissions. It needs to be highly durable. |

## 3.4.5 Transmission elements

In a variety of mechanical equipment, it is necessary to transmit power, rotary movement and speed variation between axes. To perform this function, transmission elements are used, which can be rigid or elastic. Their choice depends on the characteristics required by the system, such as: transmission ratio and speed, distance between axles, durability, cost, among others (SANTOS, 2002).

There are different types of transmission elements, the main ones being gear transmission, chain transmission and belt transmission.

### 3.4.5.1  Gear transmission

Gears are the most frequently used transmission elements, both for parallel shafts and for reverse or concurrent shafts, serving for powers, rotations and multiplication ratios that vary from minimum to maximum values. Gears are elements that can transmit forces without slipping, have a constant transmission ratio and are independent of the load, have high operational safety due to their resistance to overloads and require little maintenance (NIEMANN, 1950).

### 3.4.5.2  Chain drive

Transmissions using chains are mainly used when a greater distance between parallel axles is required. They can have ratios of up to 6 or, in extreme cases, up to 10, have an efficiency of 97 to 98% and are slip-free. This type of element can drive several wheels using a single chain and allows for larger diameters and wheelbases, but has a short service life due to wear on the joints. They are normally built to withstand power outputs of 5,000 HP, tangential forces of 28,000 kgf and speeds of 5,000 RPM (NIEMANN, 1950).

### 3.4.5.3 Belt drives

This element can be used for both parallel and reverse shafts. Its main characteristics are extremely simple construction, quiet operation and considerable capacity to absorb shocks elastically. It has a high efficiency of around 95 to 98% and a low price of around 63% compared to gears. On the other hand, their dimensions are larger, they have a short service life and there is slippage of 1 to 3% in the transmission of force (NIEMANN, 1950).

# 4. MATERIALS AND METHODS

The development of equipment such as a universal mechanical testing machine involves the study of various areas of engineering related to electrical and mechanical design, material selection, structural calculations, automation, electronics and programming.

In this context, this chapter presents the criteria for selecting the materials used and the methods adopted to develop the conceptual and structural design of the drive system for the universal mechanical testing machine.

## 4.1 Drive system

In order to develop the drive system, it was necessary to determine the elements that would make up the electromechanical assembly and the manufacturing processes that would be used. Therefore, the principle of operation, mechanical resistance and wear of the components and accessories involved in this work were studied through bibliographical research in scientific articles, catalogs and *datasheets,* as well as the sizing of these elements.

Some *software* was also used to help develop, analyze and improve the mechanical projects, such as AutoCAD *(Computer Aided Design)* to make the technical drawings and the Arduino IDE to implement the control code.

### 4.1.1 Load determination

When performing mechanical tests, the specimens to be tested are subjected to load tests. Therefore, it is necessary to characterize the drive system, sizing the load required to break the specimens analyzed and establishing limiting parameters for the system. The main parameters considered for dimensioning the load were: the type of specimen and the strength limit of the material to be tested.

The specimen chosen was for the tensile test established by ASTM D 638-02a - *Standard Test Method for Tensile Properties of Plastics,* which defines five types of specimens according to the type of material to be tested, which made it possible to define the dimensions used to calculate the system's limiting load.

Another necessary consideration was the type of material that could be tested, so we tried to size the load generated by the testing machine to be able to test most *commodity* and engineering polymers, i.e. those with a tensile strength limit ($\sigma$) OF less than 200 MPa (for example, the high mechanical strength polymer composite polyamide 6.6 with 30% glass fiber has $\sigma = 165$ MPa).

Once stresses of up to 200 MPa have been generated in the specimens, the load (F) that the system must exert can be determined using Equation 4.1.

$$\sigma = F/A \qquad \text{(Eq.4.1)}$$

where $\sigma$ is the tensile strength of the material and A is the cross-sectional area of the useful region of the specimen.

## 4.1.2 Torque transmission

The transmission ratio (í) will be used to transmit and multiply the system's torque. Attention must be paid to the change in speed, as the technical standards lay down test speed ranges depending on the material to be tested, so the transmission in the system must comply with them.

It is possible to obtain i (Equation 4.2) through the relationship between the torque generated by the system *($M_{t1}$), which comes* from the power screw, discussed in item 4.1.2.1, and the torque output from the motor shaft ($M_{t2}$), item 4.1.2.2.

$$i = Mt_1 / Mt_2 \qquad \text{(Eq. 4.2)}$$

## 4.1.2.1 Power screw

In the drive system, the power screw was used to convert torsional moment into linear displacement and generate the load for the test. The screw (Figure 5) was donated by project collaborator Prof. M.Sc. Eng. Reinaldo J. Tonete.

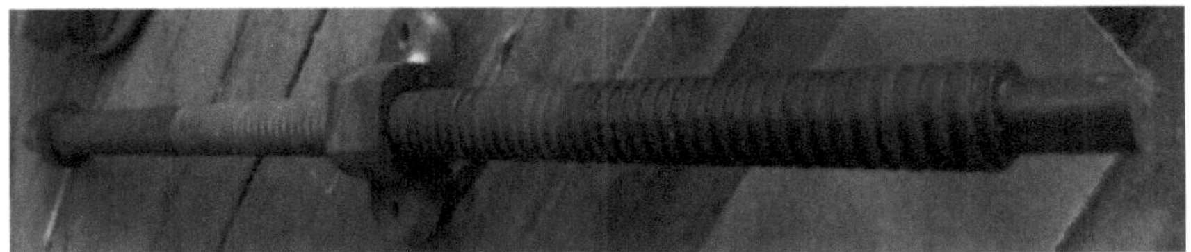

Figure 5: Power screw. Source: Oliveira (2014).

A centesimal caliper was used to obtain the dimensions of the power screw: external and internal diameter of the thread, width of the thread fillet, thread pitch, length of the thread and total length of the power screw.

Equation 4.3 was used to calculate the torsional moment generated by the power screw in dynamic motion according to the limiting load of the system and the characteristics of the screw.

$$M_t = F\, r_m\, tg(\alpha \pm \mu) = \frac{F\, r_m\, h \pm \mu\, 2\, \pi\, r_m}{2\, \pi\, r_m \mp h\, \mu} \qquad \text{(Eq. 4.3)}$$

where $M_t$ is the twisting moment, F is the load generated, $r_m$ is the average radius of the screw, a is the helix angle of the thread fillet, $\mu$ is the coefficient of friction and h is the thread pitch (NIEMANN, 1950).

## 4.1.2.1.1 Attaching the power screw to the mobile platform

To fix the power screw to the mobile platform, a fixing plate was welded to the upper end of the power screw and fixed to the mobile platform with screws. The raw material used to manufacture the fixing plate was 10 mm thick steel plate, coated electrode welding and hexagonal screws with an external diameter of 1/2 in.

The power screw had a bar extension at its upper end, which had to be cut off in order to attach it to the fixing plate. A hole was drilled in the middle of the fixing plate with a diameter equal to that of the screw's bar extension in order to fit the power screw into the hole in the fixing plate and weld them together.

Figure 6 illustrates the fixing plate which, in addition to the hole in the center for fixing the power screw, has holes at the ends to fix it to the mobile platform.

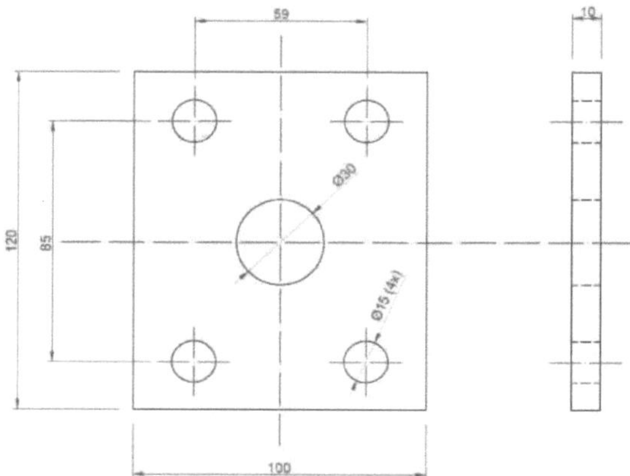

Figure 6: Power screw fixing plate. Source: Author.

Figure 7 shows the clamping plate/power screw assembly and Figure 8 shows the clamping plate/power screw assembly fixed to the mobile platform of the testing machine with threaded holes drilled in the platform.

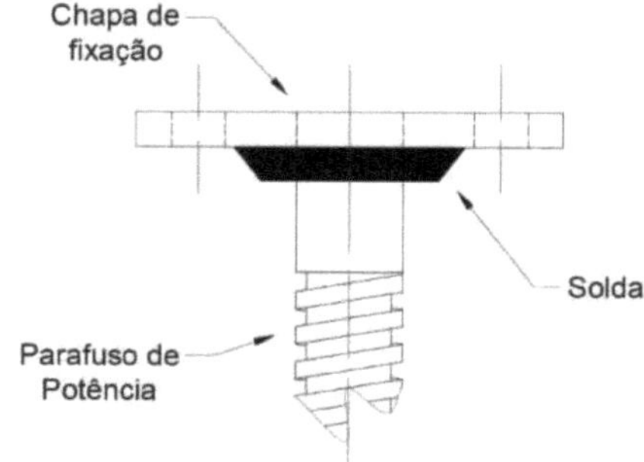


Figure 7: Power screw/plate assembly. Source: Author.

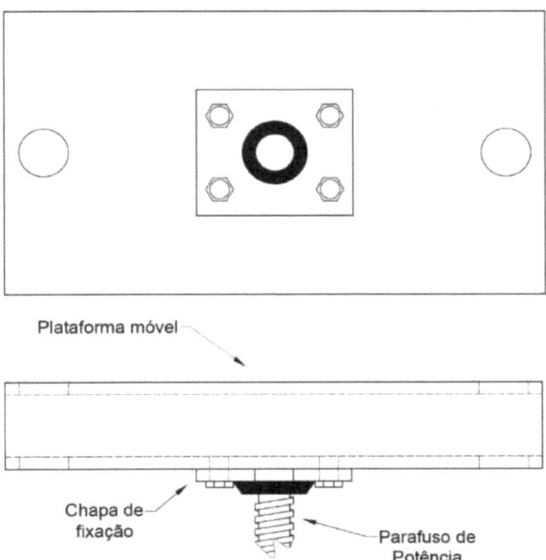


Figure 8: Power screw/fixing plate/mobile platform assembly. Source: Author.

## 4.1.2.2 Electric motor

The drive system must have an electric motor to provide rotational movement for the tests to be carried out. There are several models available on the market, so selection criteria were used to choose the electric motor so that the device meets the requirements of the system developed, such as: torque and speed variation, input power, interaction with the arduino platform, cost/benefit and availability.

Consulting the *datasheet* provided much of the information needed to select the electric motor, as well as its dimensions. With the information on the motor's torque acquired, Equation 2 was used to determine the transmission ratio required for the system.

## 4.1.2.3 System transmission element

The drive system must contain a transmission element to multiply the torque. We therefore chose the element with the right transmission ratio for the proposed system, providing the required torque. Among the elements available, it was proposed to use the pinion, chain and crown transmission set.

The crown had to be fixed so that the assembly could transmit the torque to the power screw. The hole in the pinion shaft had to be adapted to the output shaft of the electric motor and a fixing bushing developed. Figure 9 illustrates the bushing for fixing the pinion to the motor shaft.

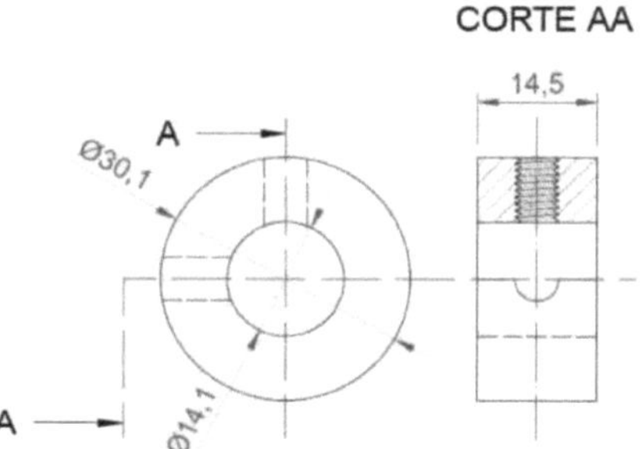


Figure 9: Bushing for fixing the pinion to the motor shaft. Source: Author.

## 4.1.2.4 Choice of bearings

The structure of the testing machine was developed so that loads of up to 4 tons could be generated, so the bearings chosen to make up the assembly must withstand the dynamic and static radial and axial loads that the system will generate.

## 4.1.2.4.1    Plates for coupling the bearings

To attach the bearings to the machine frame, 10 mm thick steel plates were used. These plates were sized according to the length and width of the boxes so that the machine would maintain its structural standard.

The plates were in direct contact with the casings, requiring milling of the faces in contact. In order to keep the bearings fixed and fitted in the equipment, a recess was made in the center of each plate with a depth of 7 mm and a diameter equal to the outer diameter of the bearing, as shown in Figure 10.

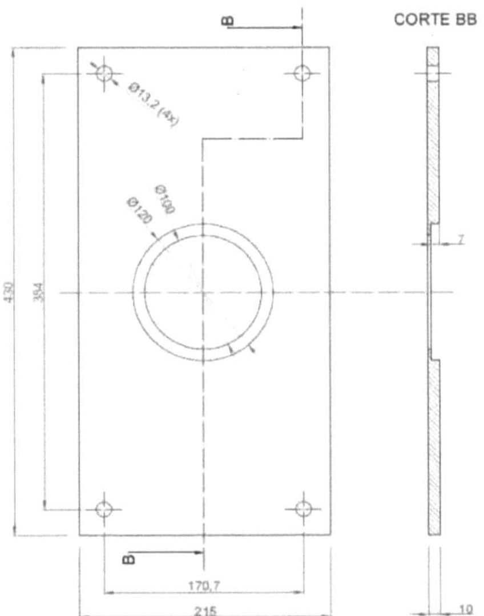


Figure 10: Bearing coupling plate. Source: Author.

In the centers, holes had to be drilled for the power screw to pass through the plates and the frames. In the plates, the holes were drilled with a diameter of 100 mm and in the boxes, the holes were drilled with 32 mm.

## 4.1.2.5 Part for fixing the crown and bearings and transmitting torque

Using the transmission assembly, a part will be developed to attach the crown and bearings to the body of the testing machine, with an internal thread to transmit the torque to the power screw.

To design it, it was necessary to know the type of material to be used for its manufacture and the number of thread fillets required to support and transmit the load generated by the system. The minimum number of thread fillets required was calculated using Equation 4.4.

$$H = \frac{3 \times F \times \cos\theta}{nt \times d_0 \times \sigma_{esc}}$$

(Eq. 4.4)

where H is the height of the thread, F is the load to be transmitted, 0 is the helix angle of the thread, nt is the number of threads working, do is the internal diameter of the thread and $\sigma_{esc}$ is the yield strength of the material.

To select the material, criteria such as: low coefficient of friction with the 1045 steel of the power screw, ease of machining, mechanical strength, cost and availability for purchase were taken into account. The diameter required for fixing the crown was the criterion for choosing the minimum diameter of the billet.

The designed part is illustrated in Figure 11, showing the dimensions needed to attach the crown, bearings and thread for the power screw.

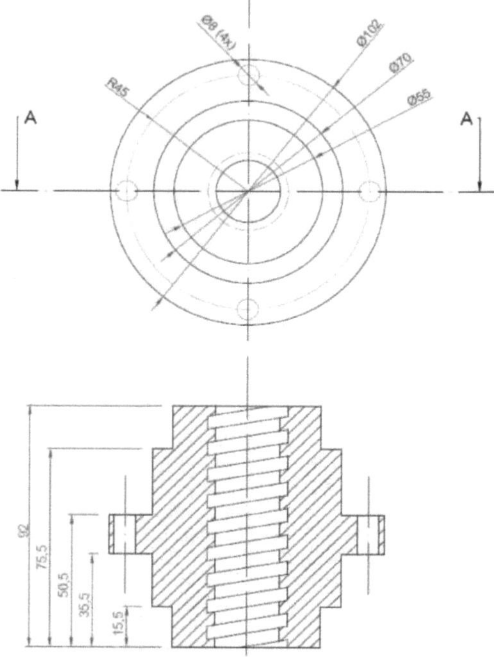

Figure 11: Part for attaching the crown/roller and for transmitting torque. Source: Author.

### 4.1.3 Rods for supporting and separating the plates

To help support and separate the plates and mechanical components of the transmission system, rods were developed for the system. The height of the rod was determined from the space needed to fit the mechanical components of the system, i.e. the height is the distance between the two bearing plates after the mechanical system has been assembled.

To attach the rods to the system, threaded rods were made to fix them by drilling through holes in the plates and sheets of the boxes. Figure 12 illustrates the rod developed for the system.

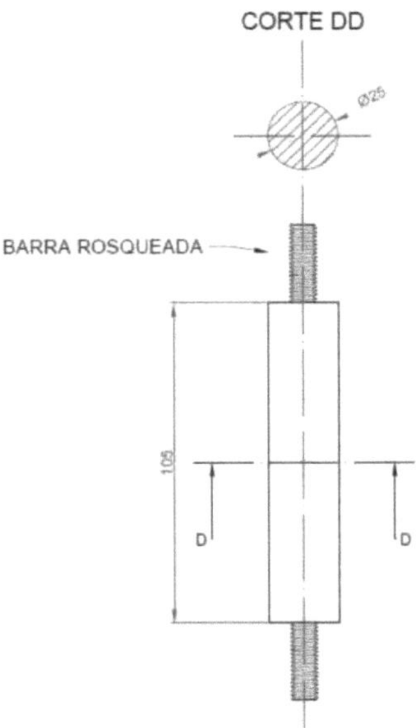


Figure 12: Rod for supporting and separating the plates. Source: Author.

### 4.1.4 Mechanical system assembly

The mechanical assembly developed is illustrated in Figure 13, which shows the components of the transmission system and the fasteners.

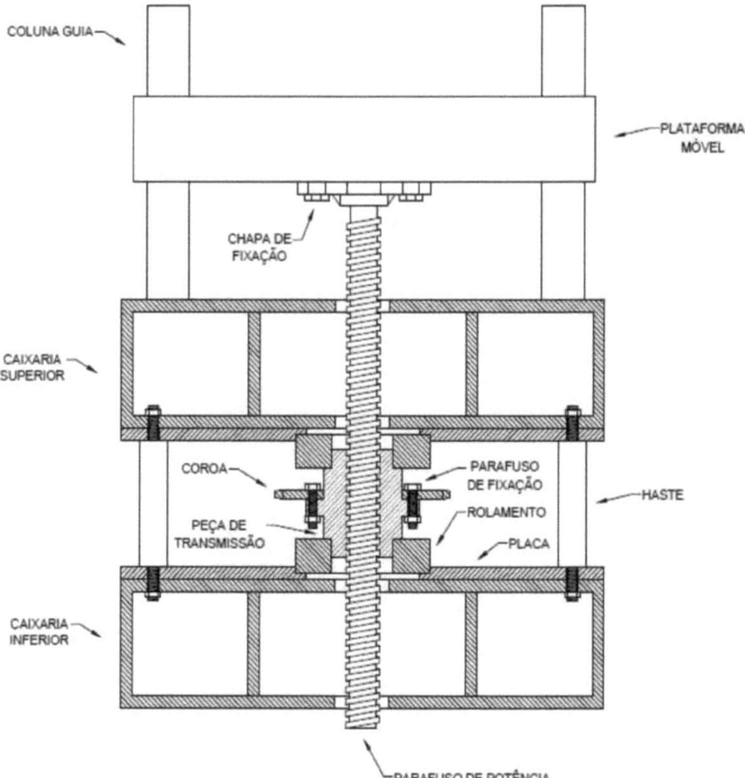


Figure 13: Mechanical assembly of the system. Source: Author.

## 4.1.5 Engine support

In order to attach the stepper motor to the machine's structure, a bracket was developed to make it possible:

- Adjusting the drive chain and

- The alignment between axles.

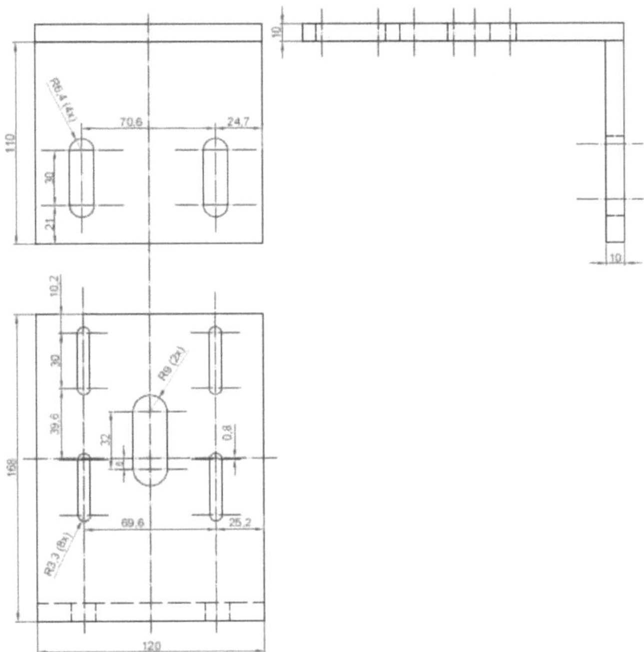

Figure 14: Motor support. Source: Author.

Figure 14 illustrates the electric motor support with the positioning and dimensions of the slots made according to the needs of the drive system.

To adjust the drive chain, 4 slots had to be made in the sheet metal with a diameter large enough to use the through bolts for fixing the motor. The holes were drilled with diameters of 6.6 mm, with a distance of 30 mm between the centers of the holes.

A slot was made in the center of the plate so that the motor shaft could be moved according to the adjustment required. The slot was made with a drilling diameter of 16 mm and a center-to-center distance of 32 mm.

To align the axes and fix the support to the machine frame, 2 slots were made in the bottom plate. The slots were drilled with a diameter of 12.8 mm and a distance of 30 mm between the centers for opening the slots.

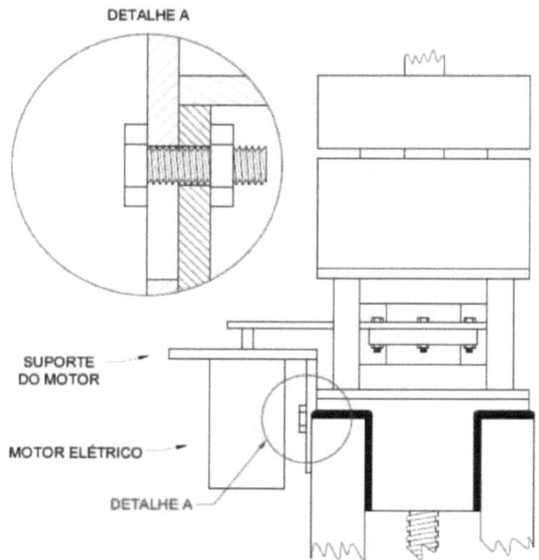


Figure 15: Support fixed to the machine frame. Source: Author.

A welding process was used to attach the motor support to the machine frame. Figure 15 shows the motor fixed to the machine frame, with a detail showing the through hole in the casing.

## 4.1.5 Machine support

The testing machine needed a support to keep the structure suspended, allowing the power screw to move freely. The support for the testing machine, illustrated in Figure 16, was made from 2" angle brackets, using the welding process to join the angle bracket pieces.

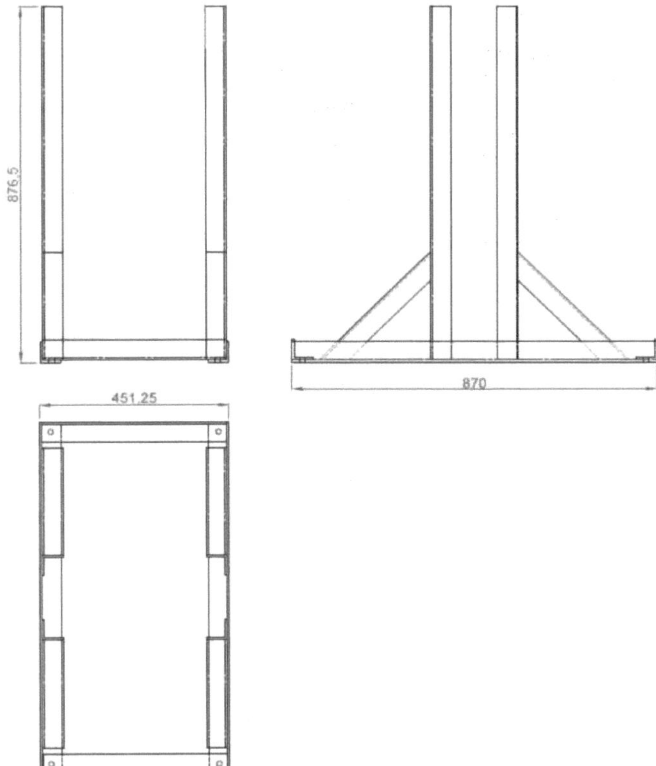

Figure 16: Testing machine support. Source: Author.

Figure 17 shows the assembly drawing of the mechanical testing machine and the support fixed to the machine frame.

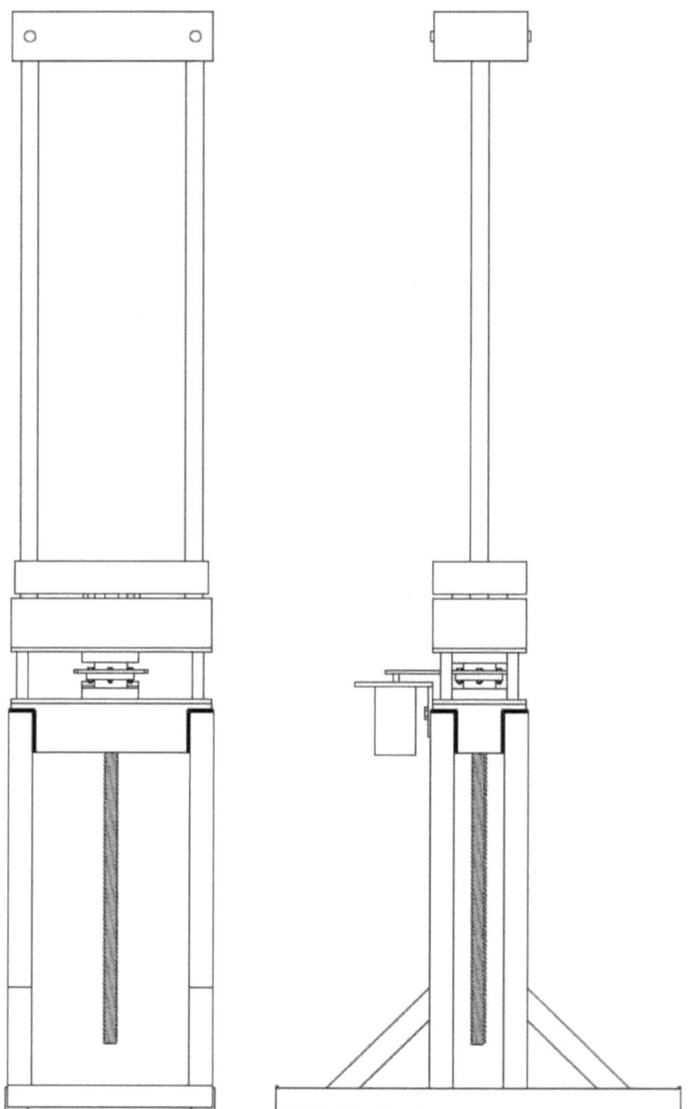

Figure 17: Universal Mechanical Testing Machine. Source: Author.

To attach it to the machine frame, a coated electrode welding process was used on the upper ends of the support, joining them to the machine's lower casing.

## 4.1.6 Control system

A drive device was used to move, control and power the bipolar stepper motor in order to obtain good performance, control the clockwise and anti-clockwise movement of the motor shaft and enable tensile and compression tests to be carried out.

The HY-DIV268N-5A driver for stepper motors (Figure 18) was used to perform these functions. This driver is compatible with 4-, 5-, 6- and 8-wire stepper motors that consume up to 5 amps per phase and require a power supply of less than 48V.

Figure 18: HY-DIV268N-5A driver. Source: Author.

Control was carried out by the Arduino Uno (Figure 19), using IDE *software* to implement the driver's programming language. The arduino was powered using the *Universal Serial Bus* (USB) connection, connected to the microcomputer. To power the driver and, consequently, the stepper motor, a switching power supply with sufficient amperage to drive the motor was used.

Figure 19: Arduino UNO. Source: Generation Robots.

The system must follow the wiring diagram illustrated in Figure 20, respecting the color sequence of the stepper motor wiring as indicated, allowing the arduino, driver, stepper motor and power supply to be connected using *jumpers*.

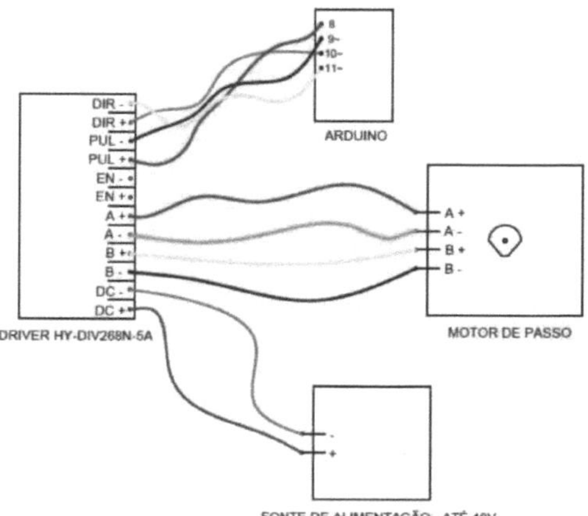


Figure 20: Arduino/driver/source/stepper motor connection diagram. Source: Author.

During the development of the control code, it was necessary to determine the distance that the mobile platform would travel during the tests. The maximum space that the platform will travel as a function of the length of the power screw was determined.

Equation 4.5 was used to determine the number of revolutions of the power screw to move the mobile platform.

$$N_{rot.} = \frac{\Delta S}{A}$$ (Eq. 4.5)

where $N_{rot.}$ is the number of revolutions of the power screw, $\Delta S$ is the space to be moved by the platform and A is the advance of the power screw.

Equation 4.6 was used to calculate the feed rate of the power screw used in the system.

$$A = P \times N_{ent.}$$ (Eq. 4.6)

where A is the screw feed, P is the pitch and Nent. is the number of inputs.

After determining the displacement that the mobile platform will travel, the number of revolutions to cover this space and the transmission ratio used in the drive system, Equation 4.7 was used to determine the number of revolutions of the stepper motor.

$$N_{rot.m.} = N_{rot.p} \times i$$ (Eq. 4.7)

where Nrot.m. is the number of revolutions of the stepper motor, Nrot.p is the number of revolutions of the power screw and $i$ is the system's transmission ratio.

The maximum number of revolutions of the stepper motor shaft was determined from the maximum distance that the mobile platform would travel, helping to create the parameters needed to control the drive system using the arduino IDE.

# 5. RESULTS

## 5.1 Drive system

The study of models of mechanical testing machines enabled us to understand the principle of operation and how the technologies involved influence the equipment, thus enabling the conceptual design of this work to be drawn up.

There are two models of testing machines on the market which differ in their operating principle for generating the test load. It was proposed to develop a drive system for an electromechanical testing machine capable of generating a load for testing polymeric specimens.

To generate the load, a stepper motor was used, which has the capacity to generate enough load to be used in the system and also allows high speed and displacement control using a computer via the arduino platform.

Electromechanical machines use two power screws in their system to transmit load to the test. However, in this work, only one power screw was used to perform this function because, based on mathematical analysis, it was realized that a single power screw is capable of transmitting sufficient load to test polymeric specimens.

### 5.1.1 Load generated by the system

In order to determine the load that the drive system must provide to the mechanical testing machine, the type of specimen that can be used for the tests must be known, as the stress must be distributed evenly over the cross-section of the specimen (GERE, 2003).

The ASTM D 638-02a - *Standard Test Method for Tensile Properties of Plastics,* standardizes five types of specimens for tensile testing for different polymeric materials, Types I, II, III, IV and V. The specimen chosen to measure the load was the Type I specimen, which has the following characteristics, Figure 21.

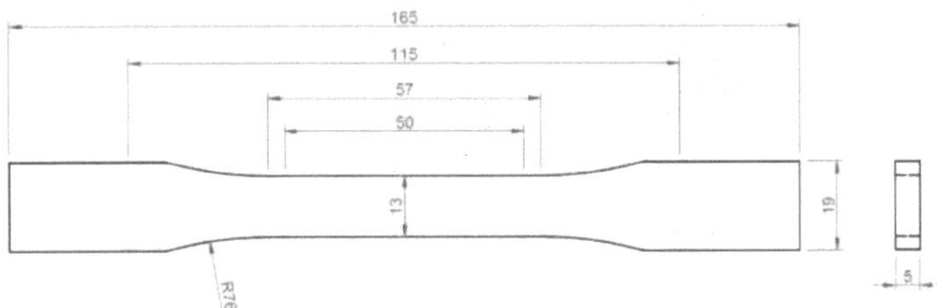


Figure 21: Type I specimen. Source: Author.

Type I specimens are widely used in tensile tests of thermoplastic, rigid, semi-rigid and non-rigid materials (ASTM STANDARDS: D 638 - 02a).

The resistance limit corresponds to the maximum conventional stress that the material can withstand without breaking (SOUZA, 1982). The tensile strength limit of 200 MPa was used for load sizing, as it allows the testing machine to generate loads to cover the largest number of *commodity* and engineering polymers that have a lower strength limit than the parameter used.

With the stress values and the dimensions of the specimen (cross-sectional area of the useful region), it was possible to dimension the maximum load for the system using Equation 4.1.

The drive system of the universal mechanical testing machine can provide a maximum test load of 13,000 N, allowing you to test

materials with a strength limit of up to 200 MPa, dimensioned according to the type I specimen.

## 5.1.2 Fixing the power screw to the mobile platform

In order for the system to perform the vertical movement for the tests, a fixing plate was made so that the power screw is fixed to the machine's mobile platform, transmitting the linear movement to the platform, consequently enabling the tests to be carried out.

It was therefore decided not to attach the power screw directly to the mobile platform that had already been milled using the welding process, so that there would be no problems with misalignment.

It was planned to use 1020 steel plate with a thickness of 10 mm for manufacture, but due to the unavailability of raw materials, the part was manufactured with plate with a thickness of 17.5 mm, and the other specifications were maintained. Four 1/2" hexagonal screws with a length of 31.5 mm were used to fix the mounting plate to the mobile platform. Figure 22 shows the fixing plate and the screws to be used to fix it to the mobile platform.

a)                              b)                              c)

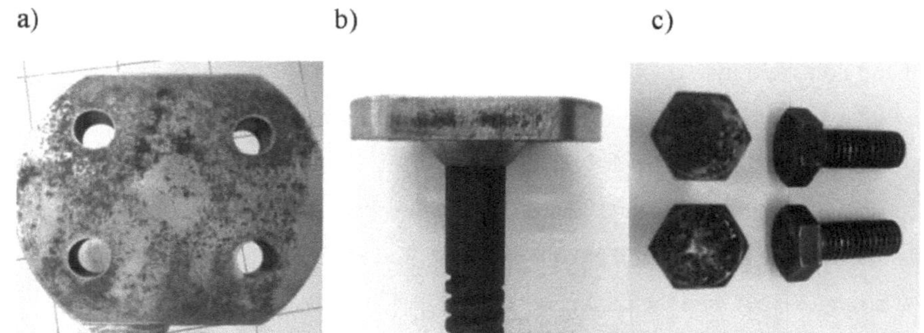

Figure 22: Fixing plate: (a) top view and (b) side view. (c) screws for fixing the plate to the mobile platform.
Source: Author.

Four 1/2" threaded holes were drilled in the mobile platform to attach the fixing plate, as shown in Figure 23.

Figure 23: Mobile platform with holes for attaching the mounting plate. Source: Author.

The milling operations carried out on the surface of the clamping plate and the mobile platform kept the machine's structure aligned, as any misalignment could compromise the movement of the mobile platform, causing the bushings to jam in the guide column, interrupting the test and potentially damaging any mechanical components in the system. Figure 24 shows the power screw/fixing plate/movable platform assembly.

a)                                               b)

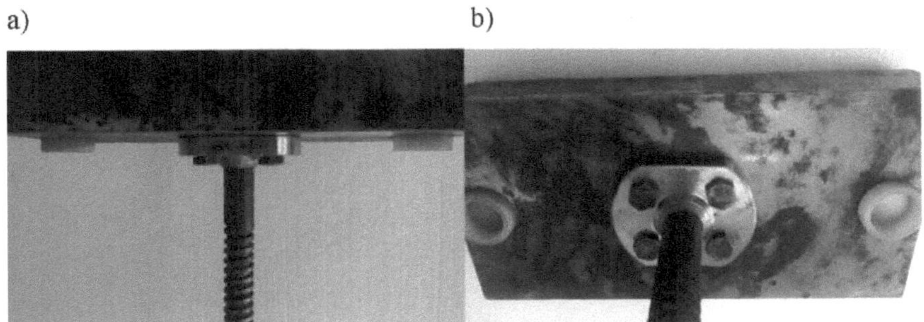

Figure 24: Power screw/fixing plate/mobile platform assembly. (a) side view and (b) bottom view. Source: Author.

Keeping the testing machine in good working order depends on the upkeep and maintenance of the equipment, so it is necessary to keep the faces that have been milled lubricated to protect them from corrosion.

### 5.1.3 Selecting the electric motor

The drive system must have a motor with precise movement and rotation speed in order to provide control over the drive and test speed, allowing control of the different speeds required by the technical standards.

Among the types of electric motors available on the market, the stepper motor has shown itself to be a device with great potential for use in the drive system, as it has high speed control and output torque, it can be controlled using the arduino platform, it has an excellent relationship between size and torque (torque proportional to the length of the body), it meets the system's requirements (to generate a load to carry out the tests with polymeric materials) and it has an excellent cost-benefit ratio.

The electric motor chosen (Figure 25) to make up the system is a 4-wire, 2-phase bipolar NEMA 37 stepper motor that generates a maximum static torque of 100kgf.cm with a current of 5 amps/phase and a voltage of 2.8V.

a)        b)

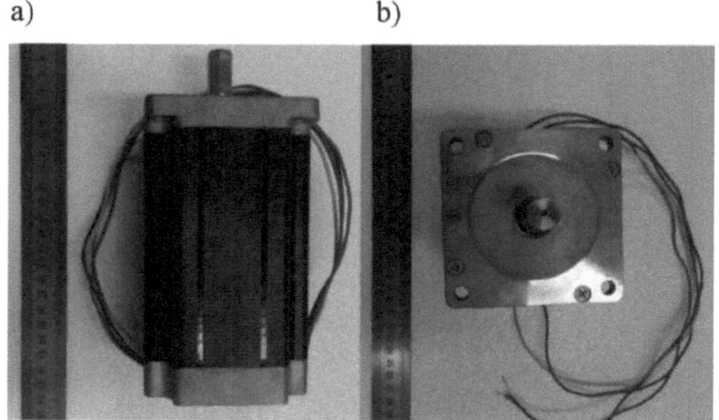

Figure 25: NEMA 37 stepper motor. (a) side view and (b) top view. Source: Author.

This motor works with a step angle of 1.8° and required a driver that worked as an H-Bridge, making it possible to change the direction of the current so that the shaft maintains rotation in the same direction or reverses the direction of 50

rotation of the shaft *(Datasheet* NEMA 37). To make one revolution the stepper motor needs to travel a specific amount of step according to the step angle, so the NEMA 37 stepper motor needs:

$$360° / 1.8° = 200 \text{ steps/revolution}$$

Knowing the number of steps will help control the output speed of the motor shaft and implement the code in the IDE.

It is not typical to use this type of motor to generate high loads, but among the types of stepper motors, the bipolar motor is the most robust and generates the highest torques among the motors of this type, around 40% more than the unipolar stepper motor.

There are standards that determine speed ranges for carrying out mechanical tests depending on the type of material, specimen and product to be tested, so using this electric motor provides high control of the drive system, allowing tests to be carried out to the specifications required by the technical standards.

## 5.1.2 System transmission element

The stepper motor chosen does not have enough torque to provide a load for the test, so in order to increase the system's torque to generate a greater load for the test, the necessary transmission ratio and the element that best suits the system were determined.

For the power screw to convert 1.3 tons into a test load, a torque of 215.74kgf.cm is required, obtained from Equation 4.3. However, the NEMA 37 stepper motor only provides a torque of 100kgf.cm, as seen in section 5.1.3.

This way, based on the torque generated by the system, coming from the power screw, and the torque output from the stepper motor, it is possible to obtain the necessary transmission ratio between the two elements so that tests can be carried out with the pre-established load on the drive system using Equation 4.2.

The system must use an element with a transmission ratio of 2.1574 in order for the system to generate the load needed to carry out the tests. Motorcycle transmission assemblies have a transmission ratio close to 3, varying according to the motorcycle model.

To make up the mechanical system, we used a transmission assembly (pinion, chain and crown) that is used in a *Biz* model motorcycle, which has a transmission ratio of 2.5 to multiply and transmit the torque to the other mechanical elements of the system, thus enabling the test to be carried out. Chain transmission has major advantages over the other two main transmission assemblies (gears and belts), such as:

- Suitable for long wheelbases (unlike gears);
- Transmit greater power (compared to belts);
- Allow length variation by removing/adding links;
- Flexibility in operating conditions;
- Transmit load with an efficiency of 97% to 98%;
- Few problems with slippage and axle alignment.

The pinion, a component of the transmission assembly, is drilled to fit the motorcycle engine shaft according to the specific vehicle model. In order to use it as the system's transmission element, it had to be adapted to the diameter of the stepper motor's output shaft, as the pinion's shaft hole had a gap of 0.07mm in relation to the motor's output shaft. To do this, a fixing bush was made with a diameter equal to the diameter of the stepper motor shaft.

The bushing had to be locked to enable the height of the pinion to be adjusted. The motor shaft is faceted on two sides, so a headless *allen* key screw was used to lock the bushing. The pinion was fixed to one side of the bushing by welding. Figure 26 shows the pinion fixed to the bushing.

Figure 26: Pinion/bushing assembly: (a) top view, (b) side view and (c) bottom view. Source: Author.

## 5.1.5 Bearings

The structure of the universal mechanical testing machine was designed and manufactured to withstand test loads of up to 4 tons in order to carry out tests using specimens made from polymeric materials.

Tapered bearings are the most suitable for this type of function, as they can withstand both axial and radial loads. Given that the transmission system generates these two types of loads, these bearings would be the most suitable.

However, considering the factors of availability and cost, the 6311 2Z/C3 ball bearing model was selected, which can withstand dynamic radial loads of 83,000 N and static loads of 47,500 N and a fatigue limit load of 3,150 N at a nominal speed of 7,100 RPM to make up the mechanical system for driving the machine. This component will help move and displace the power screw by reducing friction.

## 5.1.3 Part for coupling and torque transmission

To develop the coupling and torque transmission part, three specifications were determined regarding the dimensions it should meet.

The first specification for the coupling and torque transmission part was the internal thread. Using a centesimal caliper and a ruler, the dimensions of the power screw had to be measured, the values of which were used to make the internal thread of the coupling part. Table 3 shows the characteristics of the power screw:

Table 3: Power screw data.

| Power screw size | Value |
|---|---|
| Internal thread diameter | 24.40 mm |
| External thread diameter | 29.70 mm |
| Average thread diameter | 27.05 mm |
| Thread width | 4.40 mm |
| Thread height | 5.0 mm |
| Thread fillet helix angle | 7,5° |
| Thread pitch | 10.0 mm |
| Thread length | 910.0 mm |
| Total length of power screw | 1000.0 mm |

The second specification is the transmission assembly, as the crown has an internal bore and holes with specific diameters to fix it to the motorcycle assembly. The diameter of the internal bore determined the diameter of the recess to fit it into the transmission part and the crown fixing holes determined the exact locations of the holes in the part to fix the crown to the transmission part. The internal diameter of the crown is 70 mm and there are 4 holes for fixing it with 3/8" bolts.

The third specification refers to the bearing used in the mechanical system. The diameter of the inner ring of the bearing determined the diameter of the second recess to fit the bearings into the transmission part and the height of the bearing determined the height of the recess. The model 6311 2Z/C3 bearing was used to make up the system, with an outer diameter of 55 mm and a height of 30 mm. The bearing was not fully seated, but was partially seated in the coupling piece.

To manufacture the coupling and torque transmission part, a brass billet with a diameter of 4 in. and a height of 93 mm was used as the raw material. Brass was chosen because, in contact with steel, it has a coefficient of friction with lubrication of 0.15 - 0.19 and, above all, among the raw materials available, it was the most cost-effective.

After determining the raw material and the dimensional specifications that the part had to meet, the number of thread fillets needed to support and transmit the maximum load generated by the test machine's drive system was determined using Equation 4.4. The equation took into account $F_{system}=13,000N$, $0 = 7.5°$, $H = 3$ mm and $d0 = 27.05$ mm. The shear yield strength for brass was 70 MPa (HIBBELER, 2010).

The minimum number of threads required in the coupling part is nt = 6.8 threads. The height of the billet, 92 mm, allowed the part to be manufactured with 9.2 internal thread fillets. Figure 27 shows the part after the machining process.

Figure 27: Part for coupling crown/roller. Source: Author.

The part is designed to couple and fix the crown, bearings and interact with the power screw. Figure 28 shows the assembled crown/roller/bearing part for coupling and torque transmission.

Figure 28: Assembled set. Source: Author.

## 5.1.4 Plates for coupling the bearings

The bearings used in the mechanical system had to remain fixed to the structure of the test machine, allowing the drive system to function properly. For fixing, milled and recessed plates were used to fit the bearings.

The faces of the plates that were in direct contact with the lower and upper casings were milled to reduce misalignment and maintain greater contact between the two parts, in order to maintain the alignment of the mechanical structure of the testing machine, allowing the drive system to transmit the load efficiently.

The 120 mm diameter of the recess was made from the diameter of the outer ring of the 6311 2Z/C3 bearing so that it could be fitted into the plate with interference, securing it to the structure. The depth of the recess was set at 7 mm so that the bearing would maintain greater contact with the plate. Smaller depths could cause the bearing to loosen due to the lack of contact with the plate, as the end of the outer ring has a curvature that reduces this contact. Figure 29 shows the plates for coupling the bearings.

Figure 29: Bearing coupling plate. Source: Author.

A hole was drilled in the center of the plate with a diameter of 100 mm so that the power screw could pass through the plates, allowing it to move. The power screw has an external diameter of 29.70 mm, but the hole was drilled with a diameter of 100 mm so that the plate would not impair the operation of the bearings by locking the internal protection ring.

The 4 through holes at the ends of the plates were drilled to use 1/2" bolts for the support rods and separation of the plates. Holes were also drilled in the upper and lower frames with a diameter of 33 mm so that the power screw would pass through the machine's mechanical structure.

## 5.1.5 Support rods and plate separation

The rods were planned and designed so that they were made from a solid material with a diameter of 1", and that their ends were made in the form of a 1/2" threaded bar to fix them to the machine's mechanical assembly with through holes in the equipment.

However, because the boxes are partially closed, it was not possible to manufacture the rods as planned, as there would be no way to use any kind of wrench to adjust the rod threads.

Therefore, 4 rods were made with the same specifications as before, from solid steel with a diameter of 1", but instead of making threaded rods from the material itself, threads were opened inside the rods for 1/2" screws, as shown in Figure 30.

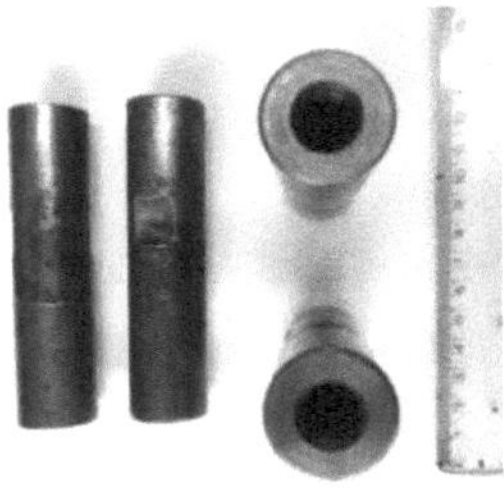

Figure 30: Fabricated rod for supporting and separating the plates. Source: Author.

Recesses were made in the body of the rods to use a 22 mm socket wrench to help fix and adjust the rods in the machine's mechanical assembly.

To attach the upper casing to the shaft, 4 hexagonal screws with 1/2" threads and 150 mm in length were used to thread into the shaft and attach the mechanical components between the bearing coupling plates on the machine frame, as shown in Figure 31.

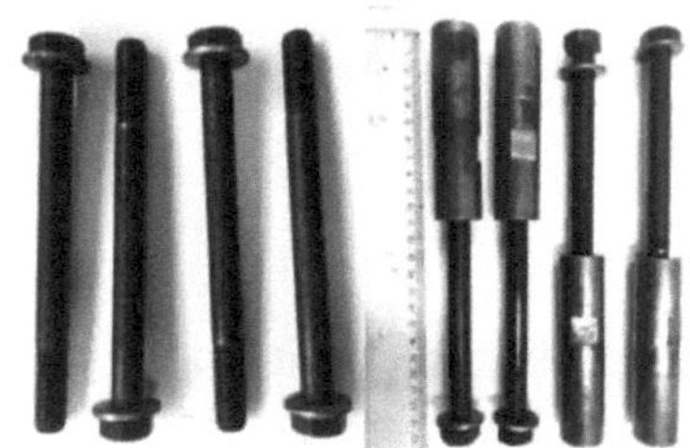

Figure 31: Screws to help fix the rods. Source: Author

Four holes were drilled in the housing so that the bolt could pass through the upper housing and the bearing coupling plate (Figure 32). Four 1/2" bolts were welded into the lower housing to thread the rods, as shown in Figure 33.

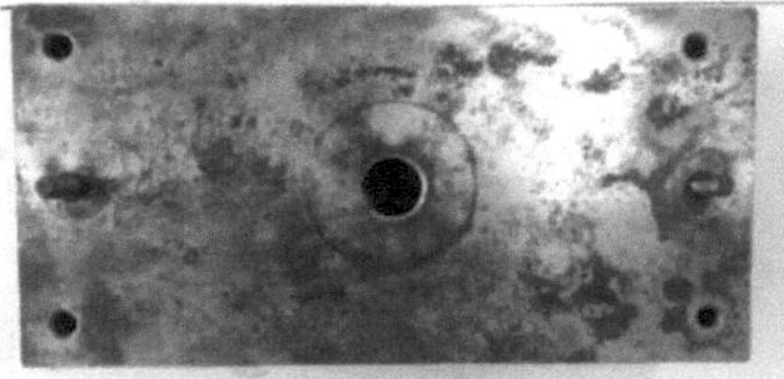

Figure 32: Upper casing with through holes. Source: Author.

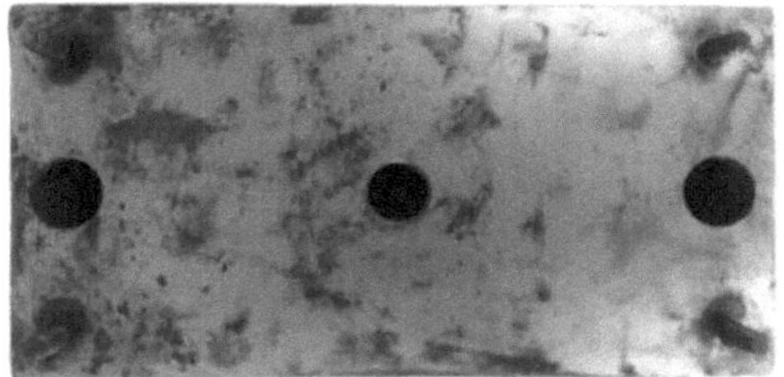

Figure 33: Lower casing with welded screws. Source: Author.

## 5.1.6 Engine support

The use of the transmission assembly made up of the sprocket, chain and crown has given the system flexibility in terms of regulation and adjustment. In order for these advantages of the assembly to be used by the system, it was necessary to develop a support that would allow the motor to be fixed to the machine frame, the chain to be adjusted and the axes to be aligned.

It is possible to add or remove links from the chain, allowing the distance between the axles to be adjusted, but it is possible that the chain does not fit completely. The motor support has therefore been designed to remove any slack that could occur when chain links are added or removed.

To attach the motor, two parameters were followed during its development: adjusting the chain and aligning the axes. The support was made from 10 mm thick steel plate in an "L" shape, with one side for attaching the motor and adjusting the chain and the other side for attaching the assembly to the machine frame and aligning the axes.

The top face of the stepper motor, where its rotor shaft is located, has four holes for fixing it to a structure. To attach the stepper motor to the bracket, the diameters of the holes in the motor were used. However, the holes in the support were made in the form of slots to allow the motor to move in a horizontal direction, promoting proper tensioning of the chain.

Once the chain has been adjusted, the axes must be aligned so as not to damage the chain or any of the system's mechanical components. The side of the support, which will be in contact with the machine's lower frame, was used to attach it to the testing machine's structure. Two 1/2" diameter slots were made for through bolts, allowing the motor/support assembly to move in a vertical direction, making it possible to align the axes of the stepper motor and the transmission part. Figure 34 shows the stepper motor support.

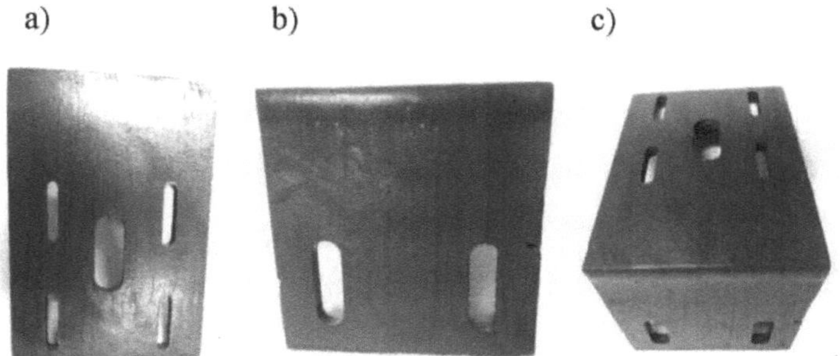

Figure 34: Motor bracket: (a) face for attaching the motor, (b) face for attaching the bracket to the machine and (c) motor bracket. Source: Author.

## 5.1.10  Machine support

The power screw used to transmit the load to the mobile platform has a total length of 1 m. When the platform is at the end of its travel (lowest travel height), the bolt extends 640 mm out of the frame.

As a result, the universal mechanical testing machine needed a support to keep its structure suspended off the ground in order to provide enough height for the power screw to move freely without hitting the ground, compromising the tests and the operation of the equipment, and to support the weight of the machine's structure.

Figure 35 shows the support for the testing machine attached to the bottom box.

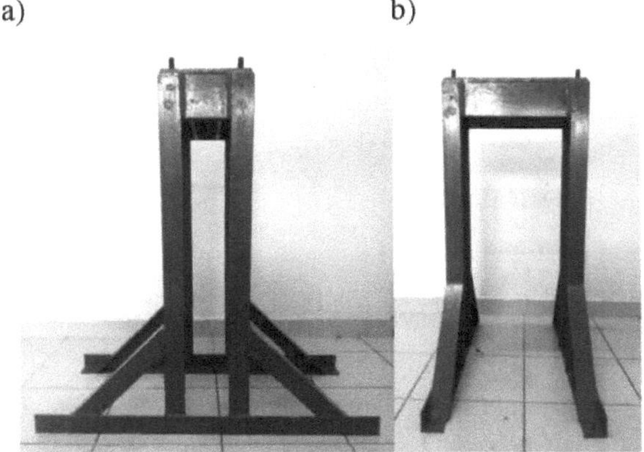


Figure 35: Machine support: (a) side view and (b) front view. Source: Author.

## 5.1.11 Control system

To use the circuit, it was necessary to identify the wires for each coil of the stepper motor so that the connection could be made according to the planned circuit. Figure 36 shows the assembled test machine control system, containing: NEMA 37 stepper motor, HY-DIV268N-5A drive, arduino UNO, power supply and the controller unit (to use the IDE *software*).

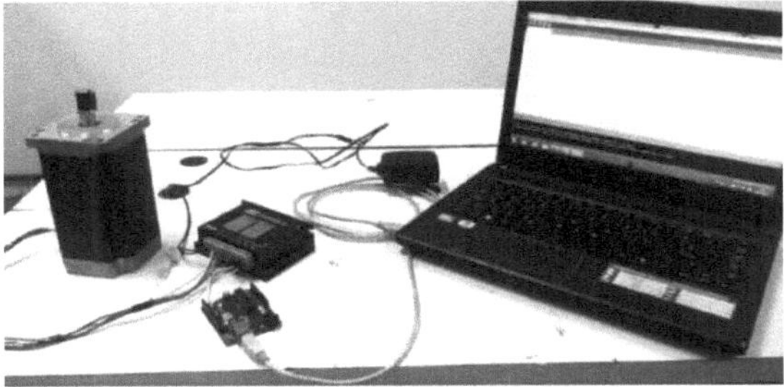

Figure 36: Control system. Source: Author.

The HY-DIV268N-5A driver has 3 switches to regulate the output current to power the stepper motor and can be set between 0.2 and 5 amps. To connect the driver, the minimum input current must be 1 ampere, so to test the circuit, a source with an output of 12 V and 1.5 A was used to power the driver and, consequently, the stepper motor.

It is necessary to use a supply with an output current of 5 A to obtain maximum performance from the stepper motor. Even using a power supply with an amperage that does not provide maximum performance, the stepper motor proved to be very efficient when using the HY-DIV268N-5A driver.

On the underside of the driver, there are fins to exchange heat and keep the device from heating up to the point of damaging it. With a power supply of 1.5 amps, the driver heated up over time, as it needed to be used long enough to adjust the code.

The arduino IDE has support for stepper motors and provides its own library for implementing the stepper motor control code. The control code implemented in the IDE *software*, Figure 37, was adapted from the source code provided by Arduino.

```
#include <Stepper.h>

const int stepsPerRevolution = 200;
Stepper Stepper(stepsPerRevolution, 8,9,10,11);

int stepCount = 0;

void setup(){
  Stepper.setSpeed(450);
  Serial.begin(9600);
}

void loop(){
  Stepper.step(200);
  Serial.print("steps:" );
  Serial.println(stepCount);
  stepCount++;

}
```

Figure 37: IDE software code implementation for controlling the NEMA 37 stepper motor.
Source: Author's own data.

The control system must comply with the speed ranges determined by the technical standards (ASTM, ABNT, DIN, among others) in order to carry out the mechanical test according to the type of material to be analyzed.

The system must be able to control speeds, so it was necessary to adapt the standard code to control the speed of the stepper motor shaft by setting the rotation speed, the number of steps to be made for the motor shaft to rotate 360° and the type of step: *full-step, half-step, 1/4 step* etc.

The type of step chosen was *full-step,* which has the highest output torque determined by the variable *"stepsPerRevolution",* so the code used 200 steps to make one revolution.

In addition to the code determining the type of step, the driver can also make this change, as the device has 3 switches that are used as a control to determine the type of step the motor should make, providing *full-step* and subdivisions of up to 1/16 of the step angle. Subdivisions of the pitch provide greater rotational speed, but result in a drop in torque. The *"setSpeed"* variable was used to determine the speed in RPM of the motor shaft regardless of the type of motor step.

The positioning of the motor shaft must be determined in the control code, as the displacement of the mobile platform is derived from the rotational displacement in the motor which is converted into linear movement by the power screw.

To implement the code, it was necessary to determine the distance the mobile platform would travel. Figure 38 illustrates the maximum displacement that the mobile platform can travel as a function of the length of the power screw simulated using *AutoCAD.*

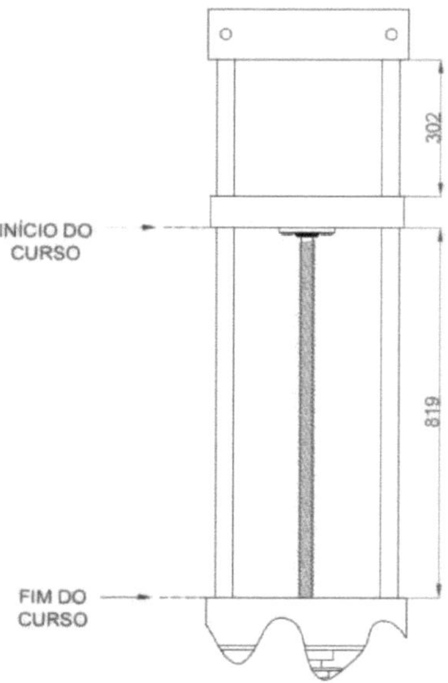


Figure 38: Maximum travel of the mobile platform. Source: Author.

The power screw has a specific number of revolutions for the platform to move. The number of rotations of the screw is the ratio between the space to be covered and the screw's feed rate, whereby the maximum displacement of the mobile platform, as a function of the length of the screw, is 819 mm and the screw's feed rate, calculated from Equation 4.6, is 10 mm. The number of revolutions of the screw for this displacement, calculated using Equation 4.5, is 81.9 revolutions.

The number of motor revolutions to achieve this maximum displacement was calculated using Equation 4.7. Thus, the stepper motor makes 245.7 revolutions to move the mobile platform over a space of 819 mm. For implementation in the code, 49,140 steps were required using the *"step"* variable, which is the function that moves the motor shaft by a specified number of steps.

Implementing the control code in the IDE made it possible to determine the positioning of the motor shaft, determining the displacement of the mobile platform so that there is no damage to the machine's structure.

## 5.1.12　Analysis of the load generated by the drive system

The structure of the universal mechanical testing machine was designed to withstand loads of up to 4 tons, but in order to test type I specimens made from polymeric materials with a strength limit of up to $CT_{resis\ t.} = 200MPa$, it was determined that loads of up to 1.3 tons should be generated by the drive system, as seen in section 5.1.1.

The stepper motor used in the system provides, at its maximum performance, a twisting moment of 100kgf.cm. When using the transmission assembly (pinion, chain and crown), which has a transmission ratio of 2.5, the transmission part receives a twisting moment of up to 250 kgf.cm, calculated from Equation 4.2.

The torque of 250kgf.cm received by the torque transmission part is transmitted to the power screw and converted into a load for the mechanical tests. The drive system can generate a load, calculated using Equation 4.3, of up to 1.57 tons, i.e. the system has a safety coefficient of 1.21. This increase in load is due to the fact that a set with a transmission ratio of 2.5 is being used.

The transmission part can support a load of 1.3 tons with 6.8 internal thread fillets, but the part was made with 9.2 thread fillets. Thus, using Equation 4.4, it was found that the part can support and transmit loads of up to 1.79 tons, i.e. the part was manufactured with a safety factor of 1.38.

## 5.1.13　Testing the drive system

The system's mechanical assembly was assembled to carry out the tests and check the performance of the mechanical components used in the drive system. Figure 39 shows the universal mechanical testing machine with the drive system.

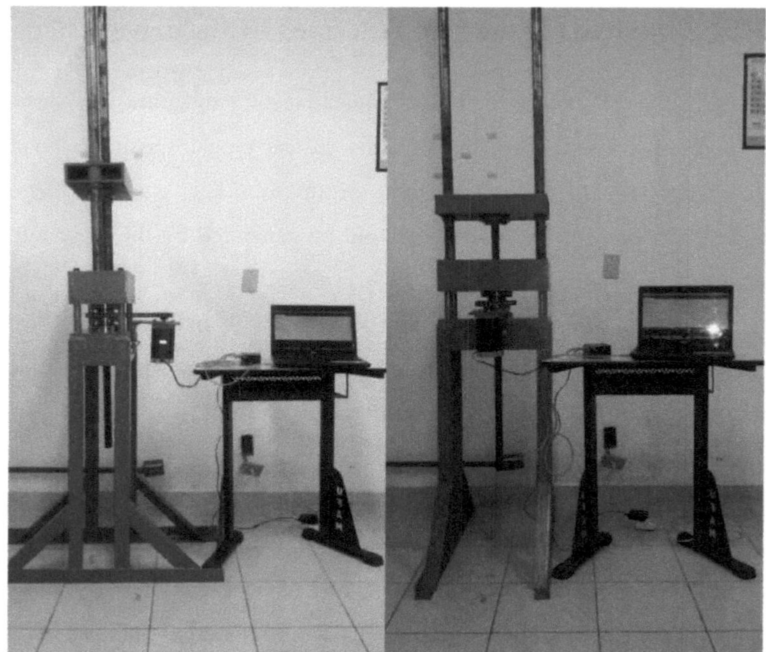
Figure 39: Drive system. Source: Author.

The tests were carried out by implementing the stepper motor control code in the arduino IDE *soft-ware*. The system used to control the stepper motor, consisting of the arduino and the HY-DIV268N-5A driver, enabled the speed and displacement of the machine's mobile platform to be controlled via the controller unit.

It was possible to change the travel speed of the mobile platform using two configuration formats:

1st: in the control code for the stepper motor, the *"setSpeed"* function was used to determine the rotation speed of the motor shaft, in RPM. In this way, the test speed can be set simply by using the IDE in the control unit. It should be noted that this function does not promote rotation, but adjusts the speed when commanded by the *"steps"* function (EVANS et al., 2013).

2°: the HY-DIV268N-5A driver has 3 switches that configure the type of step the motor makes. It is possible to use *full-step* up to subdivisions of 1/16th of a step. When setting the step in subdivisions, the motor showed a change in speed. It was therefore found that it is also possible to change the speed of rotation of the shaft using the settings in the control driver itself.

The two configuration formats differ in terms of rotation accuracy. Using the first method, it was possible to set the rotation speed accurately without changing the pitch type. The second method also changed the rotation speed, but it was not possible to determine the speed precisely.

Changing the motor's pitch type causes the output torque to change, so setting the rotation speed via the driver's switches will compromise the value of the machine's test load.

For the rotational speed settings, only the IDE was used to change it, as the pitch type was not changed, and only a 1.8° degree of rotation was used, which will help the motor reach its maximum performance of 100kgf.cm.

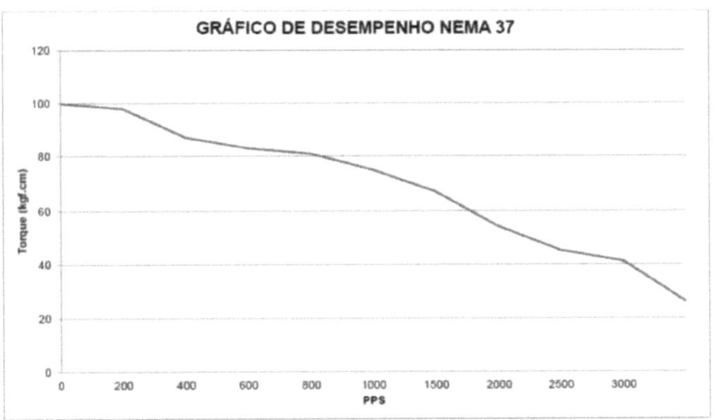


Figure 40: NEMA 37 stepper motor performance graph, torque per pulse per second (PPS) (Adapted from NEMA 37 DataSheet).

Figure 40 shows the performance graph of the NEMA 37 stepper motor. It can be seen that the torque behavior is directly related to the number of pulses per second (PPS) configured in the motor, i.e. there is a drop in the torque provided by the motor when using step subdivisions.

Therefore, in order for the system to carry out tests with loads of up to 1.3 tons, the stepper motor must provide a minimum torque of 86.3kgf.cm using up to 600 PPS (i.e. a shaft speed of 180 RPM) and the step angle can be divided into up to 3.

It should be noted that the stepper motor only starts to move when it overcomes static inertia (this can be friction, load, etc.), so the maximum rotation speed is altered by the weight of the structure, and a minimum generated load is required for the system to start working.

The *Holding Torque of* the motor is 100kgf.cm, i.e. the maximum static torque of the motor, when the speed starts to increase, the torque starts to decrease until it equals zero. It is only possible to reach the motor's maximum speed when it is rotating without being clamped to the machine.

Once the system was up and running, it was possible to see how the mechanical components were performing and, above all, to check the system's alignment.

The system showed no misalignment or warping that would hinder the movement of the mobile platform on the guide column without compromising the operation of the drive system.

Even though the system was not misaligned, it was necessary to use lubricant (grease and machine oil), which made it easier to move the platform by reducing friction between the platform bushings and the guide column and the power screw and the transmission and coupling part, resulting in smooth movement.

# CONCLUSION

The aim of this work was to develop a system to drive a universal mechanical testing machine. The conditions adopted during the development of the system limited its functionality to being able to generate a test load of 1.3 tons to test polymeric materials with a tensile strength limit of up to 200 MPa on ASTM type I specimens.

A drive system for the universal mechanical testing machine was developed using easily-acquired, simple-to-manufacture mechanical components with suitable functionality for use in the system developed. The drive system model developed differs from other systems in that it used components such as:

- A single power screw in the mechanical structure of the system. Commercial testing machines use two power screws to move the mobile platform and transmit the load.
- It has two guide columns that determine the trajectory of the mobile platform. In commercial testing machines, power screws are used to perform this function.
- The stepper motor is a device that has high speed and torque control and allows interaction with a controller unit via the arduino platform.
- The sprocket, chain and crown are used to transmit the load to the machine. Commercial machines use augers as the transmission element.

Although the drive system was designed to generate a test load of 1.3 tons, the mechanical components provided a safety coefficient of 1.38. The NEMA 37 stepper motor controlled by the HY-DIV268N-5A driver and the arduino made it possible to vary the travel speed of the mobile platform, which results in a considerable number of test speeds using the IDE *software* in the controller unit and provided a test load safety coefficient of 1.21.

The mechanical components developed made it easier to dismantle the equipment, facilitating system maintenance, allowing the machine to last longer and the components to perform better so as not to compromise the tests or damage any element.

The work sought to open up avenues for studies related to the development of systems and equipment for carrying out mechanical tests and/or the development of other drive models for the universal mechanical testing machine using different devices and technologies.

As a suggestion for future work, it is necessary to size the stepper motor taking into account the test load to be generated and the weight of the structure, as the weight of the structure compromises the maximum rotation speed. The drive system should allow greater interaction with the controller unit, generating a stress x strain graph. In this way, it will be possible to gauge the drive system by carrying out tests on polymeric materials, generating the graph from the test and comparing it with a graph of the same material generated by a commercial testing machine, verifying the behavior and discrepancy between the systems.

# REFERENCES

AKIYAMA: Technological solutions. *Stepper motor training, drivers and introduction to servo drives*. Available at: http://www.akiyama.com.br. Accessed on January 25, 2016.

ASKELAND, D. R.; PHULÉ, P. P., *Materials Science and Engineering*. 1 ed.- São Paulo: Cengage Learning, 2008.

ASTM STANDARDS: D 638 - 02a, Test Method for Tensile Properties of Plastics, 2003.

BLANCO, C., *Mechanics of Materials,* $4^a$ Edition ed. 2006: Calouste Gulbenkian Foundation.

CALLISTER, WILLIAM D. JR., *Materials Science and Engineering: An Introduction*. 5 ed. - Rio de Janeiro: LTC: Livros Técnicos e Científicos Editora S .A.

*DATASHEET HY-DIV268N-5A*. Available at: https://www.rcscomponents.kiev.ua/datasheets/div268n-5a-datasheet.pdf. Accessed on May 14, 2016.

*DATASHEETNEMA 37*. Available at: www.neoyama.com.br.

DOS SANTOS, A. A. *Cylindrical spur gears*. Faculty of Mechanical Engineering, UNICAMP. Campinas/SP, 2002.

EMIC, *Bending test*. Available at: http://www.emic.com.br. Accessed on March 31, 2016.

EVANS, M.; NOBLE, J.; HOCHENBAUM, J. *Arduino in action*. São Paulo: Novatec Editora Ltda., 2013.

GALDINO, L. *Calculation of rotation, torque and power of electric motors for power screw transmission*. Augusto Guzzo Academic Journal, 2014.

GARCIA, A.; SPIM, J. A. E DOS SANTOS, C., A. *Ensaios dos Materiais - 2$^a$ ed. -* Rio de Janeiro : LTC, 2013.

Generation Robots. Available at: https://www.generationrobots.com/blog/en/2016/09/need-help-choosing-the- right-arduino/. Accessed on January 31, 2016.

GERE, J M. *Mechanics of Materials.* São Paulo: Thomson, 2008.

GRUPO CIMM, *Universal Testing Machine,* available at: http://www.cimm.com.br/portal/material_didatico/6545-maquinas-de-ensaio-universal#.Vv0uEE_N3_g. Accessed on January 31, 2016.

HIBBELER, R.C., *Strength of Materials.* 7th ed. São Paulo: Pearson Prentice Hall, 2010.

KOYO - ROLAMENTOS DO BRASIL. Available at: http://www.koyo.com.br/. Accessed on June 17, 2016.

LEMOS, G. V. B.; VIEIRA, D. M. M.; dos SANTOS, B. P.; HAAG, J.; COSTA, V. M.; MEINHARDT, C. P.; FABRÍCIO, D. A. K.; CHIOSSI, T. P.; STROHAECKER, T. R. *The effect of strain rate on the tensile test of SAE 4340 steel.* 68th ABM Annual Congress. Belo Horizonte, Minas Gerais - BR, 2013.

MAINTENANCE AND SUPPLIES. Types of *bearings and how they work.* Available at: http://www.manutencaoesuprimentos.com.br/conteudo/3524- types-of-bearings-and-how-they-work/. Accessed on July 21, 2016.

McROBERTS M., *Arduino Basic,* Editora Novatec, São Paulo, 2011.

MONTEIRO, C., *Stepper motor - Arduino. Numerical control machines credit unit.* University of Minho. June 26, 2012.

MULTILÓGICA SHOP, *Arduino Beginner's Guide,* Available at: http://www.multilogica-shop.com. Accessed on February 19, 2016.

NIEMANN, G. *Elementos de máquinas Vol. 1*. São Paulo: Editora Edgard Blucher Ltda., 1971.

NSK Brazil. Available at: http://www.nks.com.br. Accessed on: July 15, 2016.

OLIVEIRA, J. M., *Final Technical Report - PIBIC 2013/2014*. FAPEAM, 2014.

PADILHA, A.F., *Materiais de engenharia microestrutura e propriedades*. 1st ed. Hemus AS, Brazil, 2000.

REBOUÇAS, P. P., *Development of a data acquisition system for automatic calculations of mechanical properties in a tensile test* Fortaleza: 2007

ROMERO, A. *Development of a control for stepper motors*. Department of Physics and Materials Science. São Carlos - SP, 2010.

SILVEIRA, J. A. *Experimento com Arduino,* São Paulo, 2013.

SOUZA, S. A. 1982, *Mechanical testing of metallic materials. Theoretical and practical foundations*. São Paulo: Edgard Blucher, 1982.

VIVALDINI, K. C. T. *STEPPER MOTORS*. São Carlos School of Engineering - USP. São Carlos - SP, 2009.

WERNECK, P. 2013. *Introduction to Arduino,* Available at: http://digiteducacao.net/2015/mod/page/view.php?id=31. Accessed on June 10, 2015.

ZOLIN, I. 2008. *Mechanical testing and failure analysis*. 3. ed.- Santa Maria: Federal University of Santa Maria: Industrial Technical College of Santa Maria, Industrial Automation Technical Course 2010.

# I want morebooks!

Buy your books fast and straightforward online - at one of world's fastest growing online book stores! Environmentally sound due to Print-on-Demand technologies.

Buy your books online at
**www.morebooks.shop**

Kaufen Sie Ihre Bücher schnell und unkompliziert online – auf einer der am schnellsten wachsenden Buchhandelsplattformen weltweit! Dank Print-On-Demand umwelt- und ressourcenschonend produzi ert.

Bücher schneller online kaufen
**www.morebooks.shop**

Printed by Books on Demand GmbH, Norderstedt / Germany